An Introduction to the Gas Phase

An Introduction to the Gas Phase

Claire Vallance
Department of Chemistry, University of Oxford

Morgan & Claypool Publishers

ISBN   978-1-6817-4692-0 (ebook)
ISBN   978-1-6817-4693-7 (print)
ISBN   978-1-6817-4694-4 (mobi)

DOI   10.1088/978-1-6817-4692-0

Version: 20171101

IOP Concise Physics
ISSN 2053-2571 (online)
ISSN 2054-7307 (print)

A Morgan & Claypool publication as part of IOP Concise Physics
Published by Morgan & Claypool Publishers, 1210 Fifth Avenue, Suite 250, San Rafael, CA, 94901, USA

IOP Publishing, Temple Circus, Temple Way, Bristol BS1 6HG, UK

Contents

Preface

An Introduction to the Gas Phase is adapted from a set of lecture notes for a core first year lecture course in physical chemistry taught at the University of Oxford. The book is intended to give a relatively concise introduction to the gas phase at a level suitable for any undergraduate scientist. After defining the gas phase, properties of gases such as temperature, pressure, and volume are discussed. The relationships between these properties are explained at a molecular level, and simple models are introduced that allow the various gas laws to be derived from first principles. Finally, the collisional behaviour of gases is used to explain a number of gas-phase phenomena, such as effusion, diffusion, and thermal conductivity.

Claire Vallance, September 2017

Acknowledgements

I would like to thank the first year Oxford M.Chem. students whose questions and comments over the years have directly or indirectly shaped the material covered in this book. I would also like to thank my research group for tolerating numerous writing-induced absences over the summer.

Author biography

Claire Vallance

Claire Vallance is a Professor of Physical Chemistry in the Department of Chemistry at the University of Oxford, and Tutorial Fellow in Physical Chemistry at Hertford College, Oxford. She grew up in the UK and New Zealand, and holds BSc(hons) and PhD degrees from the University of Canterbury (Christchurch, NZ). Her current research interests include chemical reaction dynamics, the use of optical microcavities in chemical sensing applications, and the development of spectroscopic techniques for use during cardiovascular surgery and neurosurgery. She has given lecture courses on chemical kinetics, properties of gases, symmetry and group theory, reaction dynamics, and astrochemistry, as well as numerous outreach and public engagement lectures, and her tutorial teaching spans the breadth of physical chemistry. She is author of over 90 journal articles, four book chapters, nine patents, an e-Textbook on symmetry and group theory, the textbooks *Astrochemistry: from the Big Bang to the Present Day*, and *An Introduction to Chemical Kinetics*, and also co-edited the textbook *Tutorials in Molecular Reaction Dynamics*.

IOP Concise Physics

An Introduction to the Gas Phase

Claire Vallance

Chapter 1

Introduction

1.1 States of matter

The gas phase is an example of a state of matter, a distinct physical form in which a substance can exist. Four states of matter are commonly encountered in our daily lives, namely solids, liquids, gases, and (less frequently) plasmas. For the rest of this book, we will limit our discussion to these four states, focusing primarily on the gas phase. However, it is worth noting that a number of more exotic states of matter can exist at extremes of temperature and pressure. These include Bose–Einstein condensates, which certain types of particle can form at very low temperature, and the quark–gluon plasma that was present at the extremely high temperatures and pressures encountered immediately after the Big Bang.

Starting from a solid at a temperature below its melting point, we can move through the four common states of matter by gradually increasing the temperature at a fixed pressure. When we reach the melting point of the solid, we overcome the bonds or intermolecular forces locking the atoms into the solid structure, and the solid melts to form a liquid. At higher temperatures virtually all of the intermolecular forces are overcome, and the liquid vapourises to form a gas[1]. If the temperature is increased still further, eventually there is enough energy to remove electrons and ionise the substance, and we form a plasma—a mixture of neutral and ionised gas and electrons. While there is much that could be (and has been) written about all of the states of matter, the focus of this book is the properties and behaviour of gases. As we shall see, the fact that interactions between gas-phase particles are only very weak allows us to use relatively simple models to gain a very good understanding of the gas phase.

---

[1] Depending on the ambient pressure and on the phase diagram of the substance (see section 1.4), it is sometimes possible for a substance to be transformed directly from the solid to the gas phase in a process known as *sublimation*.

doi:10.1088/978-1-6817-4692-0ch1   1-1

1.2 Characteristics of the gas phase

The gas phase of a substance has the following properties:

1. A gas is a collection of particles in constant, rapid, random motion, sometimes referred to as *Brownian motion*. The particles in a gas are constantly colliding with each other and with the walls of the container, with each collision changing the individual particle's direction of motion. If we followed the trajectory of a single particle within a gas, it would follow a 'random walk', an example of which is shown in figure 1.1.

2. A gas fills any container it occupies. This is a result of the second law of thermodynamics: gas expanding to fill a container is a spontaneous process due to the accompanying increase in entropy.

3. The effects of intermolecular forces in a gas are generally fairly small. For many gases over a fairly wide range of temperatures and pressures, it is a reasonable approximation to ignore them entirely. This is the basis of the *ideal gas approximation*, which we will explore in detail in section 4.1.

4. The physical state of a pure gas (as opposed to a mixture) may be defined by four physical properties:
   p - the pressure of the gas (see section 2.1);
   T - the temperature of the gas (see section 2.2);
   V - the volume of the gas, i.e. the region of space it occupies;
   n - the amount of substance present, usually expressed as a number of moles.

The physical properties (p, V, T, n) of a gas are related by an expression known, rather grandly, as an *equation of state*. If we know any three of the four properties, we can use the appropriate equation of state to calculate the fourth. In chapter 4, we will consider the equation of state for an ideal gas, in which the intermolecular forces are assumed to be zero, and we will also look briefly at some models used to describe real (i.e. interacting) gases.

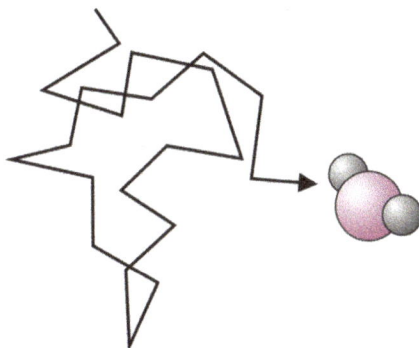

Figure 1.1. A molecule in a gas undergoes a 'random walk' trajectory as a result of multiple collisions with other gas molecules and with the walls of the container.

Elements that are gases at room temperature and atmospheric pressure are He, Ne, Ar, Kr, Xe, and Rn (atomic gases) and H_2, O_2, N_2, F_2, and Cl_2 (diatomic gases). Other substances that we commonly think of as gases include CO, NO, HCl, O_3, HCN, H_2S, CO_2, N_2O, NO_2, SO_2, NH_3, PH_3, BF_3, SF_6, CH_4, C_2H_6, C_3H_8, C_4H_{10}, and CF_2Cl_2. While these substances are all gases at room temperature and pressure, virtually every compound has a gas phase that may be accessed under the appropriate conditions of temperature and pressure. As we shall see in section 1.4, these conditions may be identified from the phase diagram for the substance.

1.3 Gases and vapours

The difference between a gas and a vapour is sometimes a source of confusion. If the gas phase of a substance is present under conditions when the substance would normally be a solid or liquid—for example, below the boiling point of the substance—then we call this a *vapour phase*. This is in contrast to a *fixed gas*, which is a gas for which no liquid or solid phase can exist at the temperature of interest. Fixed gases include substances such as N_2, O_2 or He at room temperature and atmospheric pressure.

As an example of a situation in which the vapour phase is encountered, at the surface of a liquid there always exists an equilibrium between the liquid and gas phases. At a temperature below the boiling point of the substance, the gas is in fact technically a vapour, and its pressure above the surface is known as the *vapour pressure* of the substance at that temperature. As the temperature is increased, the vapour pressure also increases. The temperature at which the vapour pressure of the substance is equal to the ambient pressure is the boiling point of the substance. We consider these ideas further in section 1.4.

1.4 Phase diagrams and phase transitions: under what conditions is a substance a gas?

A *phase diagram* predicts which state of matter a substance (or mixture of substances) will be found in under different physical conditions. The 'physical conditions' could include pressure, temperature, volume, or in the case of a mixture, composition. In the context of this primer, some familiarity with phase diagrams is useful for determining the conditions under which a particular sample will be found in the gas phase. However, if desired, readers can safely skip this section and proceed to chapter 2.

We will consider the simple case of a phase diagram for a pure substance—known as a *one component phase diagram*—plotted as a function of temperature and pressure. One-component phase diagrams for CO_2 and water are shown in figure 1.2(a) and (b), respectively. We see that in both cases, the phase diagram is divided into three areas, corresponding to the solid, liquid, and gaseous forms of the substance. By considering the effect of temperature and pressure on a substance, it is very easy to rationalise these regions. Increasing the temperature breaks intermolecular bonds, and will tend to move a substance towards the gas phase. Increasing the pressure forces molecules closer together, increasing their interactions and favouring formation of intermolecular bonds, tending to move a substance towards

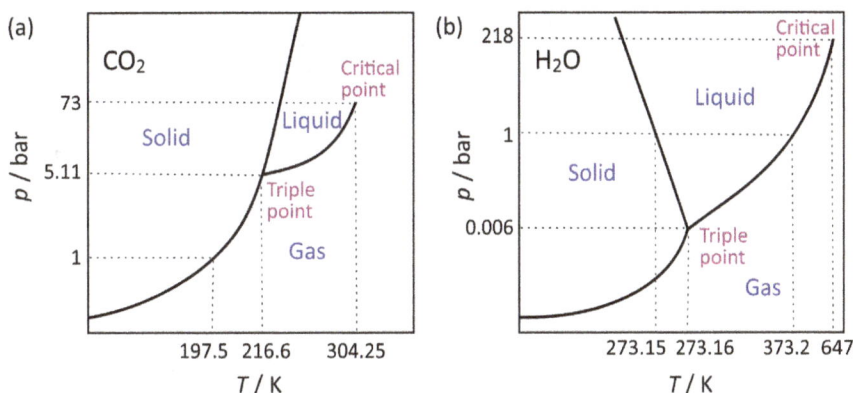

Figure 1.2. Phase diagrams of (a) CO_2 and (b) H_2O, plotted as a function of temperature and pressure.

condensed (liquid or solid) phases. We therefore find that at low temperatures and high pressures, both substances are present in solid form; at high temperatures and low pressures, both adopt the gas phase; and at intermediate temperatures and pressures, the liquid phase is thermodynamically favoured. The lines on the phase diagram denote *phase boundaries*. Crossing a phase boundary corresponds to a *phase transition*, in which a substance is transformed from one phase into another phase. At pressures and temperatures lying on a phase boundary, two phases of the substance are present at equilibrium. There is a single temperature and pressure, known as the *triple point*, at which all three phases are present at equilibrium. This is the point at which the three lines meet near the centre of the phase diagram. Finally, at high temperature we reach the *critical point*. Above this temperature, the molecules have so much energy that no amount of compression will cause gas-phase molecules to 'stick together' to form a liquid. The phase boundary between the solid and the liquid phases therefore disappears, and above the critical point there is a single fluid phase.

1.4.1 Constructing a phase diagram: the Clapeyron and Clausius–Clapeyron equations

Constructing a phase diagram and plotting the phase boundaries for a given molecule requires only some relatively simple thermodynamics, based on considering the equilibrium between two phases. At equilibrium, the Gibbs energies, G_1 and G_2, of the two phases are identical[2]. At any point on the phase boundary, for example the point labelled A in figure 1.3, we therefore have

---

[2] The change in Gibbs energy for a chemical transformation determines the direction of spontaneous change. Consider two phases with Gibbs energies G_1 and G_2, and the chemical process in which phase 1 transforms into phase 2. If $\Delta G = G_2 - G_1$ is negative, then phase 1 will spontaneously transform into phase 2, and the system is not at equilibrium. Similarly, if ΔG is positive then the reverse process becomes spontaneous: phase 2 will spontaneously transform into phase 1, and the system is also not at equilibrium. Equilibrium therefore requires that $\Delta G = 0$, i.e. that $G_1 = G_2$.

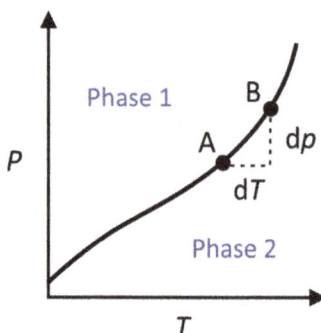

Figure 1.3. Moving an infinitesimal distance along a phase boundary between points A and B involves changing the temperature and pressure by amounts dT and dp, respectively.

$$G_1 = G_2 \tag{1.1}$$

Now consider moving along the phase boundary by an infinitesimal distance (vastly exaggerated on the diagram) to point B, which corresponds to changing the temperature and pressure by infinitesimal amounts dT and dP (note that *both* phases experience *the same* change in temperature and pressure). Since we are still on the phase boundary, the system is still at equilibrium, and so the Gibbs energies of the two phases are still equal. However, since temperature and pressure both affect the Gibbs energy, the new Gibbs energies at point B will be different from the original Gibbs energies at point A, and we have:

$$G_1' = G_2' \tag{1.2}$$

From equations (1.1) and (1.2), it is clear that the Gibbs energies of the two phases must have changed by the same amount in response to the change in temperature and pressure, i.e.

$$dG_1 = dG_2 \tag{1.3}$$

We can relate the infinitesimal changes dG_1 and dG_2 in the Gibbs energies of the two phases to the changes in temperature and pressure using a standard result from thermodynamics as follows:

$$\begin{aligned} dG_1 &= V_1\, dp - S_1\, dT \\ dG_2 &= V_2\, dp - S_2\, dT \end{aligned} \tag{1.4}$$

where V_1, V_2, S_1, and S_2 denote the volumes and entropies of phases 1 and 2, respectively. Substituting these expressions into equation (1.3) yields

$$V_1\, dp - S_1\, dT = V_2\, dp - S_2\, dT, \tag{1.5}$$

which can be rearranged to give

$$\frac{dp}{dT} = \frac{S_2 - S_1}{V_2 - V_1} = \frac{\Delta S}{\Delta V} \tag{1.6}$$

where $\Delta S = S_2 - S_1$ and $\Delta V = V_2 - V_1$ are the differences in entropy and volume between the two phases. Equation (1.6) is called the Clapeyron equation, and gives the slope of the phase boundary in terms of the changes in volume and entropy associated with the phase change. We can immediately use this equation to rationalise the slopes of the phase diagrams in figure 1.2. For example, the very steep phase boundary between the solid and liquid phases is the result of a relatively large increase in entropy due to the transformation of vibrational degrees of freedom in the solid into translational degrees of freedom in the liquid, coupled with only a small change in volume. In the case of CO_2, the volume change on going from the solid to the liquid is positive, and so is the slope of the phase boundary. Water is unusual in that the rather open, hydrogen-bonded structure of the solid results in a *reduction* in molar volume on melting, and therefore a negative slope for the corresponding phase boundary. The gradients and signs of the other phase boundaries can be rationalised in a similar way.

The entropy change ΔS associated with a phase transition can be related to the enthalpy change of transition, ΔH, and the transition temperature, T, by $\Delta S = \frac{\Delta H}{T}$. Using this relationship, the Clapeyron equation can be rewritten in the form

$$\frac{dp}{dT} = \frac{\Delta H}{T\Delta V} \tag{1.7}$$

This expression is still entirely general and can be used for any phase transition. For phase transitions between a condensed phase and a gas phase, we can make two approximations that allow us to remove the dependence on the volume change. The first approximation is that because the molar volume of a gas is much larger than that of a solid or a liquid, the change in molar volume is approximately equal to the final volume of the gas, i.e.

$$\Delta V = V_{gas} - V_{solid/liquid} \approx V_{gas} \tag{1.8}$$

The second approximation is that the gas behaves ideally, and therefore that we can use the ideal gas law, $pV = RT$, to express the volume of the gas in terms of its pressure and temperature

$$V_{gas} = \frac{RT}{p} \tag{1.9}$$

Substituting this into equation (1.7), and using the identity $(1/p)dp = d\ln p$, gives

$$\frac{d\ln p}{dT} = \frac{\Delta H}{RT^2} \tag{1.10}$$

This expression is known as the *Clausius–Clapeyron* equation. Unlike the Clapeyron equation, the Clausius–Clapeyron equation can only be used for phase transitions between a condensed phase and a gaseous phase. These two equations yield expressions for all of the phase boundaries on a phase diagram, allowing the phase diagram to be constructed when the appropriate thermodynamic data are available.

Integrating the Clausius–Clapeyron equation with respect to temperature gives an expression for the vapour pressure p of a gas above a liquid or solid at a given temperature, T.

$$\ln\left(\frac{p}{p_0}\right) = -\frac{\Delta H}{R}\left(\frac{1}{T} - \frac{1}{T_0}\right) \tag{1.11}$$

where p_0 and T_0 are standard pressure and temperature. The expressions derived above for the phase boundaries between the solid, liquid, and gas phases allow us to predict the conditions under which a substance of interest will be found in any desired phase. However, from this point onwards, we will focus on the gas phase. We have already considered the temperature and pressure of a gas in some detail in the present chapter in the context of phase diagrams. In the next chapter we will consider these properties at a more fundamental, molecular level.

Chapter 2

Pressure and temperature

2.1 Pressure

Pressure is a measure of the force exerted by a gas per unit area, and therefore has SI units of Newtons per square metre ($N\,m^{-2}$), more commonly referred to as Pascals (Pa). Several other units of pressure are in common usage, for example Torr, mmHg ('millimetres of mercury'), bar, and mbar. The relevant conversion factors are:

$$1\,Torr = 1\,mmHg = 133.3\,Pa$$
$$1\,bar = 1000\,mBar = 100\,000\,Pa$$

The 'force per unit area' exerted by a gas arises from collisions of the atoms or molecules in the gas with the surface at which the pressure is being measured, often the walls of the container[1]. Note that because the motion of the gas particles is completely random, we could place a surface at any position in a gas and at any orientation, and we would measure the same pressure.

The fact that the measured pressure arises from collisions of individual gas particles with the container walls leads directly to an important result relating to mixtures of gases, namely that the total pressure p exerted by a mixture of gases is simply the sum of the *partial pressures* p_i of the component gases. The partial pressure of one of the components i in a mixture of gases is the pressure that gas i would exert if the other gases were not present.

$$p = \sum_i p_i \qquad (2.1)$$

This result is known as Dalton's law.

---

[1] Collisions with the walls of a container are treated in detail in section 6.1.

2.1.1 Measurement of pressure

Pressure measurements are required for a wide variety of applications, spanning the range from ultra-high vacuum (10^{-10} mbar or less) up to very high pressures of many thousands of bar. There is no single physical effect that can be used to measure pressure over this vast range. However, many ingenious methods have been devised for measuring pressure over smaller ranges. At pressures higher than about 10^{-4} mbar, gauges based on mechanical phenomena may be used. These work by performing a direct measurement of the force exerted by the gas, and provide an absolute measurement, in which the determined pressure is independent of the gas species. At lower pressures, gauges tend to rely on measuring a particular physical property of the gas, which can then be related to the pressure. Such gauges must be calibrated to give correct measurements for the gas of interest. In this category, transport phenomena gauges measure gaseous drag on a moving body or exploit the thermal conductivity of the gas, while ionization gauges ionize the gas and measure the total ion current generated. The operating principles of some of the most common types of pressure gauge are outlined below.

U-tube manometer
Range: 1 mbar to atmospheric pressure
Type: mechanical

A U-tube manometer, shown in figure 2.1, consists of a U-shaped tube filled with mercury, silicon oil or some other non-volatile liquid. One end of the tube provides a reference pressure p_{ref}, and is either open to atmospheric pressure or sealed and evacuated to very low pressure. The other end of the U-tube is exposed to the system pressure p_{sys} to be measured. The gas at each end of the tube applies a force to the liquid column through collisions with the liquid surface. If the pressures at each end of the tube are unequal then these forces are unbalanced, and the liquid will move along the tube until they become balanced. This occurs when the forces exerted by the gas are matched by the force per unit area exerted by the liquid on the gas as a result of the difference in liquid height on the two sides of the U-tube. This latter force (pressure) is given by $p = \rho g h$, where ρ is the density of the liquid, g is the acceleration due to gravity, and h is the height difference between the two arms of

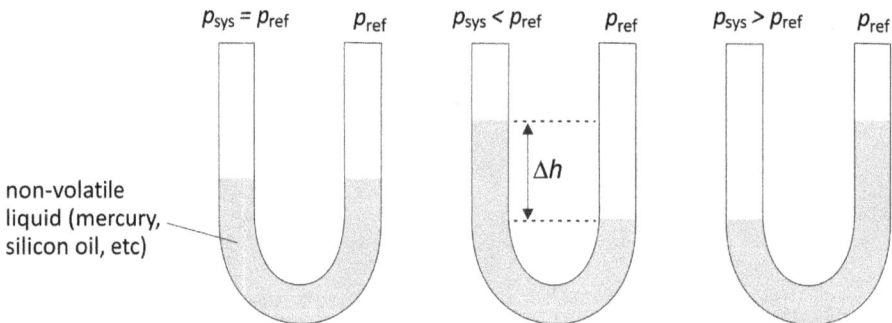

Figure 2.1. Operating principle of a U-tube manometer.

the U-tube. The balance of forces allows the unknown pressure p_{sys} to be calculated from the known reference pressure and the measured height difference of the liquid.

$$p_{sys} = p_{ref} + \rho g h \qquad (2.2)$$

Bourdon gauge
Range: 1 mbar up to high pressure (at least tens of bar)
Type: mechanical
A Bourdon gauge works on the same principle as a 'party blower'. As shown in figure 2.2(a), the gauge head contains a 'C' shaped, fine-walled, hollow metal tube known as a *Bourdon tube*. When pressurised, the cross section of the tube changes shape, and the tube flexes and attempts to straighten. The tube is connected to a gearing system that transforms the flexion of the tube into rotation of a pointer, which in turn indicates the pressure on a scale.

Capacitance manometer
Range: 10^{-6} to 10^{5} mbar
Type: mechanical
A capacitance manometer, sometimes known by the MKS trade name 'Baratron', contains a thin metal diaphragm that is deflected when the pressure changes. The deflection is sensed electronically via a change in capacitance between the diaphragm and one or more fixed electrodes. One side of the diaphragm is maintained at a reference pressure, and the measured capacitance therefore allows an absolute pressure to be determined. The arrangement is shown schematically in figure 2.2(b).

Thermocouple gauges and Pirani gauges
Range: 10^{-4} up to 1000 mbar
Type: transport
Thermocouple gauges and Pirani gauges both contain a metal wire that is heated by an electrical current. While the wire is being heated, collisions of the surrounding gas with the wire carry heat away from the wire and cool it, with the net result that the

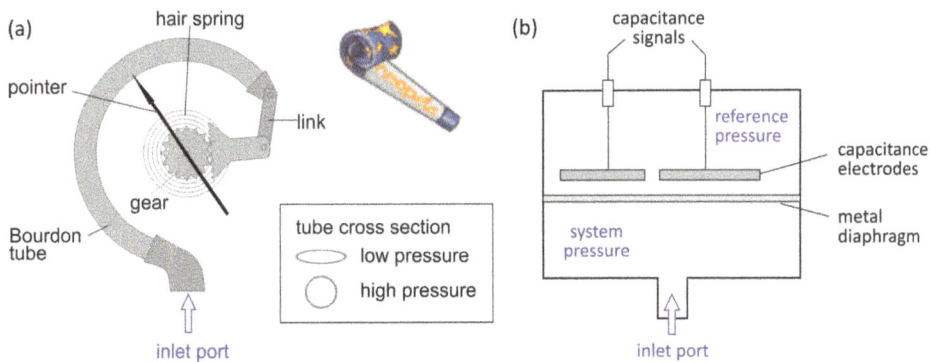

Figure 2.2. (a) Bourdon gauge; (b) capacitance manometer.

wire temperature settles at some equilibrium value determined by the current passing through the wire and the pressure of the gas. If the pressure is lowered, heat is carried away less efficiently and the temperature of the wire increases, while an increase in pressure leads to more efficient cooling and a decrease in the wire temperature. The temperature of the wire may therefore be used to measure the pressure. In a thermocouple gauge, a thermocouple is used to measure the temperature of the wire directly. In a Pirani gauge, the temperature is determined by monitoring the electrical resistance of the wire, which has a known dependence on temperature.

Hot cathode ionization gauge (Bayard–Alpert gauge)

Range: 10^{-10} to 10^{-3} bar
Type: ionization
A hot cathode gauge consists of a heated filament that emits electrons, an acceleration grid, and a thin wire detector, as shown schematically in figure 2.3. Electrons emitted from the filament are accelerated towards the grid, ionizing molecules in the surrounding gas along the way. The ions are collected at the detection wire, and the measured ion current is proportional to the gas pressure. This type of ionization gauge has the advantage that there is a linear dependence of the ion current on the gas pressure. Like any ionization gauge, correction factors need to be applied for different gases to account for differences in electron ionization probability.

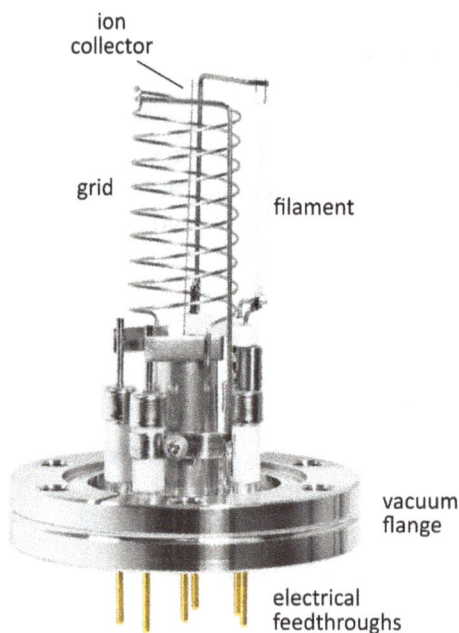

Figure 2.3. Bayard–Alpert (hot cathode) ionization gauge.

Cold cathode ionization gauge (Penning gauge)
Range: 10^{-9} to 10^{-2} mbar
Type: ionization
A cold cathode gauge works on a similar principle to a hot cathode gauge, but with a different ionization mechanism. There is no filament to produce electrons, simply a detection rod (the anode) and a cylindrical cathode, to which a high voltage ($\sim$4 kV) is applied. Ionization is initiated randomly by a cosmic ray or some other ionizing particle entering the gauge head[2]. The electrons formed are accelerated towards the anode. A magnetic field causes them to follow spiral trajectories, increasing the path length through the gas, and therefore the chance of ionizing collisions. The resulting ions are accelerated towards the cathode, where they are detected. More free electrons are emitted as the ions bombard the cathode, further increasing the signal. Eventually a steady state is reached, with the ion current being related to the background gas pressure. The relationship is not a linear one as in the case of a hot cathode gauge, and the pressure reading is only accurate to within around a factor of two. In its favour, the Penning gauge is more damage-resistant than a hot cathode gauge.

2.2 Temperature

The temperature of a gas is a measure of the amount of kinetic energy possessed by the constituent particles, and therefore reflects the velocity distribution within the gas. Whenever a particle undergoes a collision with another particle or with the walls of the container, energy and momentum can be transferred to or from the particle. If we followed the velocity of any single particle within a gas, we would see it changing rapidly with time as a result of these collisions. However, when considering the entire ensemble of particles, conservation of energy means that collisions can only lead to exchange of energy between the particles, and not to an increase or decrease in their total energy. The total number of particles with a given velocity is therefore constant, even though the identities of the particles with a particular velocity change over time through collisions. In other words, at a given temperature, the velocity distribution of the gas particles is conserved.

Note that the temperature of a substance is a direct result of the motion of its constituent atoms and molecules. This has the important consequence that the concept of 'temperature' only has any meaning in the presence of matter. It is impossible to define the temperature of a perfect vacuum, for example. In addition, temperature is only really a meaningful concept for systems at thermal equilibrium.

The distribution of molecular speeds, $P(v)$, within an ideal gas at thermal equilibrium is given by the Maxwell–Boltzmann distribution, which we will derive later on in section 5.2.

$$P(v) = 4\pi \left(\frac{m}{2\pi k_B T} \right)^{3/2} v^2 \exp\left(-\frac{mv^2}{2k_B T} \right) \tag{2.3}$$

---

[2] This occurs more frequently than you might think!

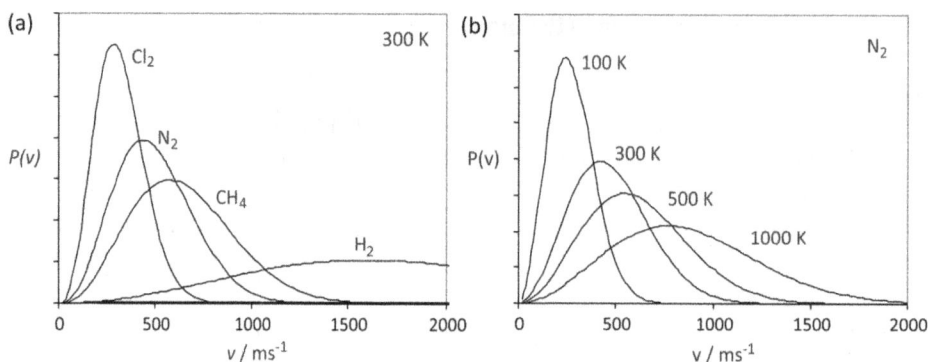

Figure 2.4. Maxwell–Boltzmann distributions for: (a) common gases of various molecular masses at a temperature of 300 K; (b) N_2 at temperatures in the range from 100 K to 1000 K.

The Maxwell–Boltzmann distribution is a function of the ratio m/T, where m is the mass of each gas particle and T is the temperature. The particle speed distribution therefore depends on both of these properties. The plots in figure 2.4 show the Maxwell–Boltzmann speed distributions for a number of different gases at two different temperatures. As we can see, average molecular speeds for common gases at room temperature (300 K) are generally a few hundred metres per second. For example, N_2 has an average speed of around 500 m s^{-1}, rising to around 850 m s^{-1} at 1000 K. A light molecule such as H_2 has a much higher mean speed of around 1800 m s^{-1} at room temperature. We can make two observations:

1. Increasing the temperature broadens the speed distribution and shifts the peak to higher speeds. This means that there are more 'fast' particles at higher temperatures, but it is important to note that there are still many 'slow' particles as well.

2. Decreasing the mass of the gas particles has the same effect as increasing the temperature, i.e., heavier particles have a slower, narrower distribution of speeds than lighter particles.

We will consider some further consequences of the Maxwell–Boltzmann distribution when we look at collisions in more detail in chapter 5. However, now that we have introduced the concept of temperature, we will consider some of the ways in which temperature can be measured.

2.2.1 Thermal equilibrium and measurement of temperature

Based on our current definition, one way to measure the temperature of a gas would be to measure the speed of each particle and then to find the appropriate value of T in the expression for the Maxwell–Boltzmann distribution (equation (5.25)) to match the measured distribution. This is clearly impractical, due both to the extremely high speeds of the gas particles and the difficulties associated with tracking any given particle amongst a sea of identical particles. Instead, temperature measurements generally rely on the process of thermal equilibration.

If two objects at different temperatures are placed in contact, heat will flow from the hotter object to the cooler object until their temperatures equalise. When the two temperatures are equal, we say that the objects are in *thermal equilibrium*. The concept of thermal equilibrium leads to the 'zeroth law' of thermodynamics: '*If A is in thermal equilibrium with B and B is in thermal equilibrium with C, then A is also in thermal equilibrium with C*'. This statement provides the basis for a rather formal definition of temperature as being 'that property which is shared by objects in thermal equilibrium with each other'. The zeroth law may seem very obvious, but it is an important principle when it comes to measuring the temperature of a system. In general, it is usually not practical to place two arbitrary systems in thermal contact to find out if they are in thermal equilibrium and therefore have the same temperature. However, the zeroth law means that we can use the properties of some reference system to establish a temperature scale, calibrate a measuring device to this reference system, and then use the device to measure the temperature of other systems. An example of such a device is a mercury thermometer. The reference system is a fixed quantity of mercury, and the physical property used to establish the temperature scale is the volume occupied by the mercury as a function of temperature. To make a temperature measurement, the mercury is allowed to come into thermal equilibrium with the system we are measuring, and the volume occupied by the mercury once equilibrium has been established may be converted to a temperature on our previously-established scale.

Standard mercury or alcohol thermometers therefore rely on the physical property of thermal expansion of a fluid for temperature measurement. However, many other properties may also be used to measure temperature. Some of these include:

1. Electrical resistance
   The resistance of an electrical conductor or semiconductor changes with temperature. Devices based on metallic conductors are usually known as *resistance temperature devices*, or RTDs, and rely on the more-or-less linear rise in resistance of a metal with increasing temperature. A second type of device is the *thermistor*, which is based on changes in resistance within a ceramic semiconductor. Unlike metallic conductors, the resistance of these devices drops non-linearly as the temperature is increased.

2. Thermoelectric effect
   When a metal is subjected to a thermal gradient, a potential difference is generated. This effect is known as the thermoelectric (or Seebeck) effect, and forms the basis for a widely-used class of temperature measurement devices known as thermocouples[3].

---

[3] A thermocouple employs a junction between two different metals, which is placed in contact with the substance to be measured. The thermoelectric effect leads to a potential difference between the ends of the two wires, which can be related to the temperature of the substance. Different pairs of metals are used to make measurements over different temperature ranges.

3. Infra-red emission

All substances emit black-body radiation with a wavelength or frequency distribution that reflects their temperature. Infra-red temperature measurement devices measure emission in the infra-red region of the spectrum in order to infer the temperature of a substance or object.

4. Thermal expansion of solids

Bimetallic temperature measurement devices consist of two strips of different metals, bonded together. The distinct thermal expansion coefficients of the two metals results in one side of the bonded strip expanding more than the other on heating, causing the strip to bend. The degree of bending provides a measure of the temperature.

5. Changes of state

Thermometers based on materials that undergo a change of state with temperature are becoming increasingly widespread. For example, liquid-crystal thermometers undergo a reversible colour change with changes in temperature. Other materials undergo irreversible changes, which may be useful in situations where all we need to know is whether a certain temperature has been exceeded, for example packaged temperature-sensitive goods.

Having considered each of the properties of a gas in isolation, in the next chapter we move on to investigating the relationships between them.

An Introduction to the Gas Phase

Claire Vallance

Chapter 3

Relationships between gas properties: the gas laws

3.1 The relationship between pressure and volume

Figure 3.1(a) illustrates the observed relationship between the volume and pressure of a gas at two different temperatures. Initially, we will focus on just one of the curves. We see that as we increase the pressure from low values, the volume of a gas first drops precipitously, and then at a much slower rate, before more or less leveling out to a constant value. In fact, we find that *pressure is inversely proportional to volume*, and that the curves follow the equation

$$pV = \text{constant} \qquad (3.1)$$

This relationship, known as *Boyle's law*, suggests that it becomes increasingly more difficult to compress a gas as we move to higher pressures. It is fairly straightforward to explain this observation using our understanding of the molecular basis of pressure. Consider the experimental setup shown in figure 3.1(b), in which a gas is compressed by depressing a plunger that forms the 'lid' of the container. When the plunger is at its highest position (shown on the left-hand side of the figure), the volume occupied by the gas is large and the pressure is low. The low pressure means that there are relatively few collisions of the gas with the inside surface of the plunger, and the force opposing depression of the plunger is correspondingly low. Under these conditions it is very easy to compress the gas. Once the plunger has already been depressed some way (as shown on the right-hand side of the figure), the gas occupies a much smaller volume, there are many more collisions with the inside surface of the plunger, and the gas has a higher pressure. The collisions generate a large force opposing further depression of the plunger, and it becomes much more difficult to reduce the volume of the gas.

doi:10.1088/978-1-6817-4692-0ch3

3-1

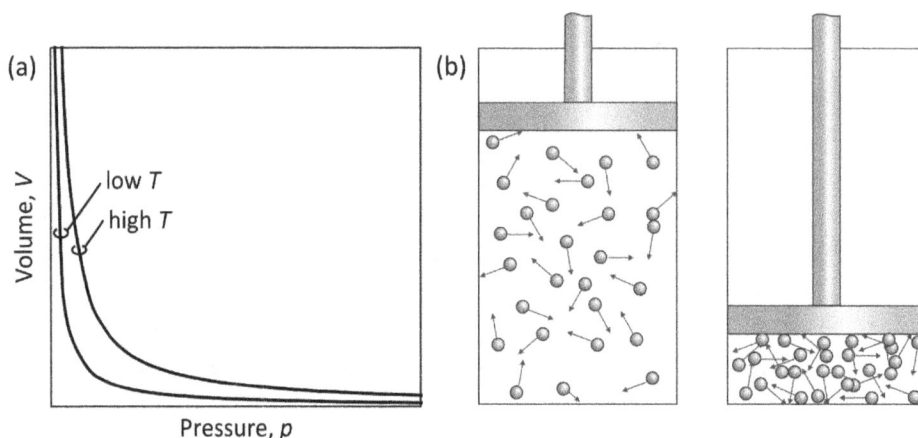

Figure 3.1. (a) Relationship between pressure and volume for a gas at two different temperatures; (b) compression of gas by a plunger leads to an increase in pressure.

3.2 The effect of temperature on pressure and volume

From the two plots shown in figure 3.1(a), we see that for a fixed volume, the pressure increases with temperature. In fact, this is a direct proportionality:

$$p \propto T \quad \text{(at constant volume)} \tag{3.2}$$

Similarly, we find that at a fixed pressure, the volume is linearly dependent on temperature.

$$V \propto T \quad \text{(at constant pressure)} \tag{3.3}$$

This second relationship is known as *Charles's law* (or sometimes *Gay-Lussac's law*). It is often written in the slightly different (but equivalent) form

$$\frac{V_1}{T_1} = \frac{V_2}{T_2} \tag{3.4}$$

where V_1 and V_2 are the volumes of the gas at temperatures T_1 and T_2, respectively. Equations (3.2) and (3.3) may be combined to give the result

$$pV \propto T \tag{3.5}$$

All of these observations are again very straightforward to explain using our molecular understanding of gases. The primary effect of increasing the temperature of the gas is to increase the speeds of the particles. As a result, at higher temperature there are more collisions with the walls of the container (or the inside surface of the plunger in our example above), and the collisions are also of higher energy. For a fixed volume of gas, these factors combine to give an increase in pressure. On the other hand, if the experiment is to be carried out at constant pressure, we require that the total force exerted upwards on the plunger through collisions remains constant. Since, on average, individual collisions are more energetic at higher temperatures, this may only be achieved by reducing the number of collisions, which requires a reduction in the density of the gas and therefore an increase in its volume.

3.3 The effect of the amount of gas, n

It follows fairly intuitively from the arguments above that since the number of collisions must be proportional to the number of gas molecules in the sample, both pressure and volume must also be proportional to this quantity, i.e.,

$$pV \propto n \qquad (3.6)$$

This relationship is known as *Avogadro's principle*.

3.4 Equation of state for an ideal gas

We can combine all of the above results into a single expression known as the *ideal gas law*,

$$pV = nRT \qquad (3.7)$$

where the constant of proportionality, R, is called the *gas constant*, and takes the value 8.314 J K^{-1} mol^{-1}. R is related to Boltzmann's constant, k_B, by $R = N_A k_B$, where N_A is Avogadro's number.

The ideal gas law is the equation of state for an ideal gas (ideal gases will be discussed in more detail in section 4.1), and is also a reasonable approximation to the equation of state for real gases. It generally provides a good description of most gases at relatively low pressures and moderate to high temperatures, the conditions under which the original experimental observations described above were made. To understand why the ideal gas law often breaks down at high pressures and low temperatures, we need to consider the differences between an ideal gas and a real gas. This is the topic of chapter 4.

IOP Concise Physics

An Introduction to the Gas Phase

Claire Vallance

Chapter 4

Ideal gases and real gases

4.1 The ideal gas model

The ideal gas model is an approximate model of a gas that is often used to simplify calculations on real gases. An ideal gas has the following properties:

1. There are no intermolecular forces between the gas particles;

2. The volume occupied by the particles is negligible compared to the volume of the container they occupy;

3. The only interactions between the particles and with the container walls are perfectly elastic collisions[1].

Of course, in a real gas, the atoms or molecules have a finite size, and at close range they interact with each other through a variety of intermolecular forces, including dipole–dipole interactions, dipole-induced dipole interactions, and van der Waal's (induced dipole–induced dipole) interactions. When applied to real gases, the ideal gas model breaks down when molecular size effects or intermolecular forces become important. This occurs under conditions of high pressure, when the molecules are forced close together and therefore interact strongly, and at low temperatures, when the molecules are moving slowly and intermolecular forces have a long time to act during a collision. The pressure at which the ideal gas model starts to break down depends on the nature and strength of the intermolecular forces between the gas particles, and therefore on their identity. The ideal gas model becomes more and more exact as the pressure is lowered, since at very low pressures the gas particles are widely spaced and interact very little with each other.

---

[1] An *elastic collision* is a collision in which the total kinetic energy is conserved, meaning that no energy is transferred from translation into rotation or vibration or vice versa, and no chemical reaction occurs.

doi:10.1088/978-1-6817-4692-0ch4

4.2 The compression factor, Z

In a real gas there are attractive and repulsive intermolecular forces between gas particles. As a result, at a given temperature and pressure, the molar volume of a real gas is likely to be different from that of an ideal gas. For example, when attractive forces dominate, the real gas will have a smaller volume than the ideal gas, as the attractive forces pull the particles closer together. The reverse is true when repulsive interactions dominate.

The deviations of a real gas from ideal gas behaviour may be quantified by a parameter known as the *compression factor*, usually given the symbol Z. The compression factor is simply defined as the ratio of the molar volume V_m of the gas to the molar volume V_m^{ideal} of an ideal gas at the same pressure and temperature.

$$Z = \frac{V_m}{V_m^{ideal}} \tag{4.1}$$

The value of Z provides information on the dominant types of intermolecular forces acting in a gas.

$Z = 1$: No intermolecular forces, ideal gas behaviour

$Z < 1$: Attractive forces dominate, and the gas occupies a smaller volume than an ideal gas.

$Z > 1$: Repulsive forces dominate, and the gas occupies a larger volume than an ideal gas.

All gases approach $Z = 1$ at very low pressures, when the average spacing between particles is large. To understand the behaviour at higher pressures we need to consider a typical intermolecular potential, $V(r)$, shown in figure 4.1(a), which describes the energy of interaction between two molecules as a function of their separation, r. We can divide the potential into three regions, as illustrated in the diagram, and consider the value of Z in each region.

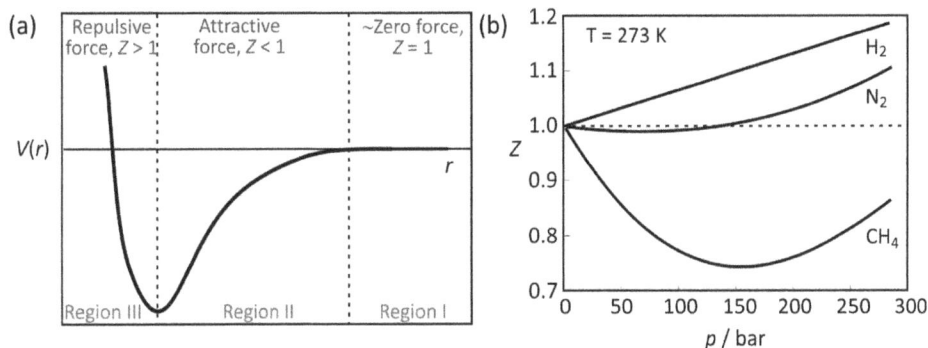

Figure 4.1. (a) A typical intermolecular potential, labelled according to the contribution of various regions of the potential to the compression factor; (b) compression factor, Z, as a function of pressure at 273 K for H_2, N_2, and CH_4.

1. *Region I: large separations*

   At large separations the interaction potential is effectively zero, and $Z = 1$. We therefore expect the gas to behave ideally, and this is indeed the case, with Z tending towards unity for all gases at sufficiently low pressures.

2. *Region II: small separations*

   As the molecules approach each other, they experience an attractive interaction, whereby the potential energy of the system is lowered (and the system is therefore stabilised) as the particles move closer together. This draws the molecules in the gas closer together than they would be in an ideal gas, reducing the molar volume such that $Z < 1$. These conditions are found at intermediate pressures.

3. *Region III: very small separations*

   At very small separations (found in high pressure systems), the electron clouds on the molecules start to overlap, giving rise to a strong repulsive force. Bringing the molecules closer together now increases their potential energy. Because of the intermolecular repulsion, the molecules now take up a larger volume than they would in an ideal gas, and $Z > 1$.

The behaviour of Z with pressure for a few common gases at a temperature of 273 K (0 °C) is illustrated in figure 4.1(b). At low pressures the compression factor is close to unity. At higher pressures Z becomes negative for most molecules as attractive interactions tend to dominate. The attractive interactions are seen to be particularly strong for CH_4, much weaker for N_2, and virtually non-existent for H_2. At still higher pressures, Z tends towards more positive values as the repulsive part of the intermolecular potential becomes important.

The compression factor also depends on temperature. The reasons for this are two-fold, but both stem from the increase in molecular speeds at higher temperature. Firstly, at higher speeds there is less time during a collision for the attractive part of the potential to act, and the effect of the attractive intermolecular forces is therefore smaller. The effect of the attractive interactions on the trajectories of a pair of

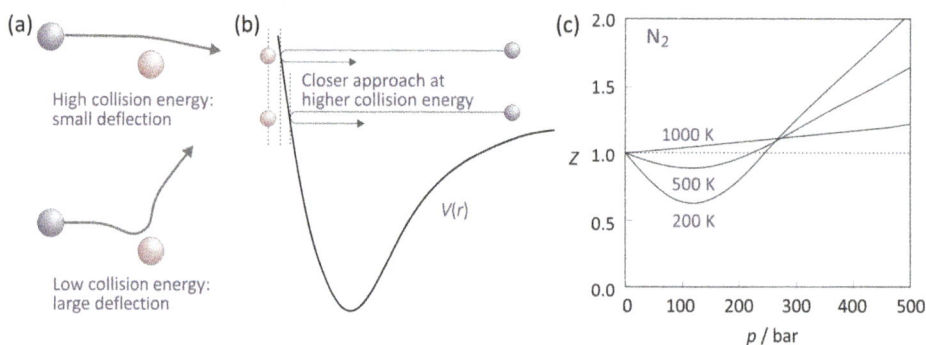

Figure 4.2. (a) Effect of collision energy (temperature) on particle trajectories during a collision; (b) higher energy collisions at higher temperatures penetrate further into the repulsive wall of the interaction potential; (c) compression factor for N_2 as a function of pressure at three different temperatures.

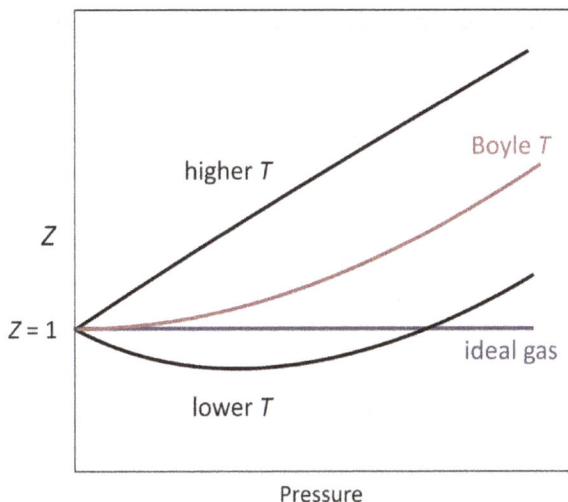

Figure 4.3. In contrast to an ideal gas, the compression factor of a real gas varies with pressure. At low pressures, the behaviour of a real gas approaches that of an ideal gas at a single temperature known as the Boyle temperature.

colliding particles at different speeds is illustrated in figure 4.2(a). Secondly, the higher energy of the collisions allows the colliding particles to penetrate further into the repulsive part of the potential during each collision, so the repulsive interactions become more dominant (see figure 4.2(b)). The temperature of the gas therefore changes the balance between the effects of attractive and repulsive interactions. As an example, the pressure-dependent compression factor for N_2 at three different temperatures is shown in figure 4.2(c), and clearly shows the effects of reduced attraction and increased repulsion as the temperature is increased.

Finally, we should consider the rate of change of Z with pressure. An ideal gas has $Z = 1$ and $dZ/dp = 0$, i.e., the slope of a plot of Z against p is zero at all pressures. For all real gases, Z tends towards unity at low pressures. However, dZ/dp only tends towards zero at a single temperature called the *Boyle temperature*, T_B. At the Boyle temperature, the attractive and repulsive interactions exactly balance each other and the real gas behaves ideally over a certain range of (low) pressures. This is shown schematically in figure 4.3.

4.3 Equations of state for real (non-ideal) gases

There are a number of ways in which the ideal gas law (equation (3.7)) may be modified to take account of the intermolecular forces present in a real gas. One way is to treat the ideal gas law as the first term in an expansion of the form:

$$pV = RT(1 + B'p + C'p^2 + \cdots) \qquad (4.2)$$

This is known as a *virial expansion*, or sometimes as the *virial equation of state*. The coefficients B', C', etc are called *virial coefficients*. The virial expansion is often seen in the alternative form:

$$pV = RT\left(1 + \frac{B}{V} + \frac{C}{V^2} + \cdots\right) \tag{4.3}$$

In many applications, only the first correction term, with coefficient B or B' (known as the *first virial coefficient*), is included. Note that the Boyle temperature mentioned above is the temperature at which the first virial coefficient is equal to zero.

Another widely used equation for treating real gases is the *van der Waals* equation,

$$p = \frac{nRT}{V - nb} - a\left(\frac{n}{V}\right)^2 \tag{4.4}$$

where a and b are temperature-independent constants called the *van der Waals coefficients*. The constant a encapsulates the effect of intermolecular interactions, while b accounts for the non-zero volume occupied by the particles in a real gas. Each gas has its own characteristic van der Waals coefficients. The van der Waals equation is often expressed in terms of molar volumes[2] V_m.

$$p = \frac{RT}{V_m - b} - \frac{a}{V_m^2} \tag{4.5}$$

---

[2] Equation (4.5) is obtained from equation (4.4) by dividing the top and bottom of each fraction by the number of moles of gas, n.

IOP Concise Physics

An Introduction to the Gas Phase

Claire Vallance

Chapter 5

A molecular perspective: the kinetic theory of gases and the molecular speed distribution

Now that the fundamental concepts determining the properties of gases have been covered, we are ready to move on to a more quantitative description. The ideal gas model, which represents a simplified approximate version of a real gas, has already been introduced. We will find in the following sections that we can use this model as the basis for developing the *kinetic theory of gases*. The name comes from the fact that within kinetic theory, it is assumed that the only contributions to the energy of a gas arise from the kinetic energies of the gas particles. Note that this is also implicit in the assumptions of the ideal gas model listed at the start of section 4.1.

Kinetic theory is a powerful model that allows us to relate macroscopic measurable quantities to motions on the molecular scale. In the following sections, we will use kinetic theory to calculate 'microscopic' quantities such as average particle velocities, collision rates, and the distance travelled between collisions, as well as to investigate macroscopic properties such as pressure and transport phenomena; for example, diffusion rates and thermal conductivity.

5.1 Collisions with the container walls—determining pressure from molecular speeds

As described in section 2.1, the measured pressure of a gas arises from collisions of the gas particles with the walls of the container. By considering these collisions more carefully, we can use kinetic theory to relate the pressure directly to the average speed of the gas particles. We begin by determining the momentum transferred to a container wall in a single collision. Figure 5.1 below shows a particle of mass m and velocity v colliding with a wall of area A. Before the collision, the particle has velocity component v_x and momentum component $p_{\text{initial}} = mv_x$ in the direction of the wall. We define this direction as the x axis for the purposes of the present derivation. After the collision, the particle has momentum $p_{\text{final}} = -mv_x$ along the

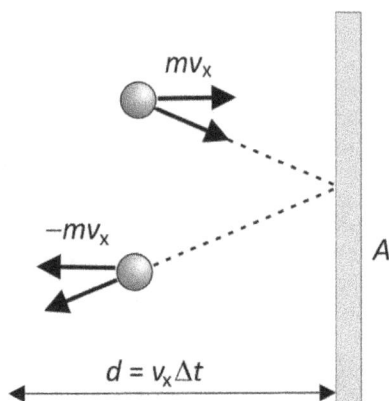

Figure 5.1. Momentum transfer on collision of a particle with one of the container walls.

x direction, with the components of momentum along y and z remaining unchanged. Since momentum must be conserved during the collision, and the momentum of the particle has changed by an amount $\Delta p = p_{\text{final}} - p_{\text{initial}} = 2mv_x$, the total momentum imparted to the wall during the collision must also be $2mv_x$.

The next step is to determine the total number of collisions with the wall occurring within a given time interval Δt. During this time interval, all particles within a distance $d = v_x \Delta t$ of the wall (and travelling towards it) will collide with the wall. Since the area of the wall is A, this means that all particles within a volume $Av_x \Delta t$ will undergo a collision. We now need to work out how many particles will be within this volume and travelling towards the wall. The number density of the molecules (i.e., the number of molecules per unit volume) is

$$\text{number density} = \frac{N}{V} = \frac{nN_A}{V} \tag{5.1}$$

where N is the number of molecules and n the number of moles in the container of volume V. The number of molecules within our volume of interest, $V = Av_x \Delta t$, is therefore just the number density multiplied by this volume, i.e.,

$$\text{number of molecules} = \frac{nN_A}{V}Av_x \Delta t \tag{5.2}$$

Since the random velocities of the particles mean that on average half of the molecules in the container will be travelling towards the wall and half away from it, the number of molecules within our volume travelling towards the wall is half of the above value. The total momentum imparted to the wall is now just the momentum change per collision multiplied by the total number of collisions,

$$\Delta p_x = 2mv_x \left(\frac{1}{2} \frac{nN_A}{V} Av_x \Delta t \right) = \frac{nMAv_x^2 \Delta t}{V} \tag{5.3}$$

where we have used $M = mN_A$, with M the molar mass.

Pressure is defined as the force per unit area, so we need to convert the above momentum into a force exerted on the surface in order to calculate the pressure. We can do this using Newton's second law of motion,

$$F_x = ma_x = m\frac{dv_x}{dt} = \frac{dp_x}{dt} \tag{5.4}$$

where a_x is the acceleration along the x direction. Applying this result to equation (5.3), we obtain

$$F_x = \frac{dp_x}{dt} = \frac{\Delta p_x}{\Delta t} = \frac{nMAv_x^2}{V} \tag{5.5}$$

The pressure is therefore

$$p = \frac{F_x}{A} = \frac{nMv_x^2}{V} \tag{5.6}$$

Finally, there is a small amount of 'tidying up' to carry out on this expression. We have based our arguments on a particle with a single velocity, v_x. However, in reality there is a distribution of velocities within the gas, and we should replace v_x^2 with $\langle v_x^2 \rangle$, the average of this quantity over the distribution. We can simplify matters still further by recognising that the random motion of the particles means that the average speed along the x direction is the same as along y and z. This allows us to define a root mean square speed:

$$v_{rms} = \left[\langle v_x^2 \rangle + \langle v_y^2 \rangle + \langle v_x^2 \rangle\right]^{1/2} = [3\langle v_x^2 \rangle]^{1/2} \tag{5.7}$$

Rearranging this expression, we find

$$\langle v_x^2 \rangle = \frac{1}{3}v_{rms}^2 \tag{5.8}$$

Substituting this into equation (5.6) gives

$$p = \frac{1}{3}\frac{nMv_{rms}^2}{V} \tag{5.9}$$

or equivalently, multiplying both sides through by V,

$$pV = \frac{1}{3}nMv_{rms}^2 \tag{5.10}$$

We have performed this final step to show that, since the average speed of the molecules is constant at constant temperature, our simple treatment of collisions with a surface has enabled us to derive Boyle's law, i.e.,

$$pV = \text{constant} \quad \text{(at constant temperature)} \tag{5.11}$$

From here, it is fairly straightforward to go one step further and derive the ideal gas law. The equipartition theorem (see appendix) states that each translational degree of freedom possessed by a molecule is accompanied by a contribution of $(1/2)k_BT$ to its internal energy. Kinetic theory assumes that the only available degrees of freedom for each particle are its three translational degrees of freedom, with kinetic energy $(1/2)mv_{rms}^2$. Equating these two results gives

$$\frac{3}{2}k_BT = \frac{1}{2}mv_{rms}^2 \qquad (5.12)$$

Multiplying both sides through by Avogadro's number, N_A, and rearranging slightly gives

$$RT = \frac{1}{3}Mv_{rms}^2 \qquad (5.13)$$

Finally, substituting this result into equation (5.10) yields the ideal gas law.

$$pV = nRT \qquad (5.14)$$

Our simple kinetic model of gases can therefore explain all of the experimental observations described in chapter 3.

5.2 The Maxwell–Boltzmann distribution revisited

In section 2.2, we introduced the Maxwell–Boltzmann distribution, which describes the speed distribution of gas molecules at thermal equilibrium. There are various ways in which this distribution may be derived. In the following version much of the hard work is avoided by means of fairly straightforward symmetry arguments.

We will start by breaking the velocity v down into its components v_x, v_y and v_z, and considering the probability $P(v_x)dv_x$ that a particle has a velocity component v_x in a range dv_x, i.e., the probability that the velocity component lies between v_x and $v_x + dv_x$. Since each velocity component may be treated independently, according to probability theory the total probability of finding a particle with components v_x, v_y, v_z in the range dv_x, dv_y, dv_z is just the product of the probabilities for each component.

$$P(v_x, v_y, v_z)dv_xdv_ydv_z = P(v_x)dv_xP(v_y)dv_yP(v_z)dv_z \qquad (5.15)$$

Since the directions in which the particles travel are completely random, and all directions within the gas are equivalent, we can reason that the distribution function $P(v_x, v_y, v_z)$ can actually only depend on the total speed v of the particle rather than on the individual velocity components. To express v in terms of the components we use the fact that $v^2 = v_x^2 + v_y^2 + v_z^2$. The probability distribution function[1] may now be written as $P(v_x^2 + v_y^2 + v_z^2)$, and we have

---

[1] Note that $P(v_x^2 + v_y^2 + v_z^2)$ is the probability density function. To obtain a probability of the particle having a particular set of velocity components, it must be multiplied by the volume element $dv_xdv_ydv_z$.

$$P\left(v_x^2 + v_y^2 + v_z^2\right)dv_x dv_y dv_z = P(v_x)P(v_y)P(v_z)dv_x dv_y dv_z \tag{5.16}$$

and therefore

$$P\left(v_x^2 + v_y^2 + v_z^2\right) = P(v_x)P(v_y)P(v_z) \tag{5.17}$$

The only type of function that satisfies a relationship of this kind is an exponential (using the fact that $e^{x+y+z} = e^x e^y e^z$), so we can see immediately that the functions $P(v_x)$, $P(v_y)$, and $P(v_z)$ on the right-hand side of equation (5.17) must in fact be exponential functions of $v_x{}^2$, $v_y{}^2$, and $v_z{}^2$. It is fairly easy to show that a suitable solution is:

$$P(v_x) = A\exp(-Bv_x^2) \tag{5.18}$$

where A and B are constants, with analogous expressions for $P(v_y)$ and $P(v_z)$. The argument in the exponential is negative because energy constraints mean that for our model to make physical sense, the probability of finding a particle must decrease as we go to higher particle speeds. Determining the two constants is fairly straightforward. Since $P(v_x)$ is a probability distribution, it must be normalised to unity, i.e.,

$$\begin{aligned}
1 &= \int_{-\infty}^{\infty} P(v_x)dv_x \\
&= A\int_{-\infty}^{\infty} \exp(-Bv_x^2)dv_x \\
&= A\left(\frac{\pi}{B}\right)^{1/2}
\end{aligned} \tag{5.19}$$

where we have used the standard integral $\int_{-\infty}^{\infty} e^{-ax^2}dx = (\pi/a)^{1/2}$ for $a > 0$.

From equation (5.19), we see that $A = (B/\pi)^{1/2}$. We can now determine B by using the distribution to calculate a property that we already know. From equations (5.7) and (5.13), we have

$$\langle v_x^2 \rangle = \frac{1}{3}v_{\mathrm{rms}}^2 = \frac{k_B T}{m} \tag{5.20}$$

We can also calculate $\langle v_x^2 \rangle$ using the probability distribution given in equation (5.23). The average value of a property x that can take any value in a continuous range, and has a probability $P(x)$ of taking a particular value, is given by $\langle x \rangle = \int xP(x)dx$. We therefore have

$$\langle v_x^2 \rangle = \int_{-\infty}^{\infty} v_x^2 P(v_x)dv_x = \left(\frac{B}{\pi}\right)^{1/2}\int_{-\infty}^{\infty} v_x^2 \exp(-Bv_x^2)dv_x \tag{5.21}$$

The integral is a standard integral of the same form as we met before in equation (5.19), and has the value $\frac{1}{2}(\frac{\pi}{B^3})^{1/2}$, yielding

$$\langle v_x^2 \rangle = \frac{1}{2B} \tag{5.22}$$

From equation (5.20), we therefore find that $B = m/(2k_BT)$, and the probability distribution function becomes

$$P(v_x) = \left(\frac{m}{2\pi k_BT}\right)^{1/2} \exp\left(-\frac{mv_x^2}{2k_BT}\right) \tag{5.23}$$

The final step is to use this result to determine the distribution of molecular speeds, rather than just the distribution of a single velocity component. From equation (5.15), this is simply

$$P(v_x, v_y, v_z)dv_xdv_ydv_z = \left(\frac{m}{2\pi k_BT}\right)^{3/2} \exp\left(-\frac{mv^2}{2k_BT}\right)dv_xdv_ydv_z \tag{5.24}$$

The above expression gives the probability of the speed distribution having the specific components v_x, v_y, v_z lying within particular ranges dv_x, dv_y, dv_z, whereas what we would really like to know is the probability $P(v)dv$ that the molecular speed lies in the range from v to $v + dv$. We can find this quantity simply by integrating the above distribution over the spherical shell bounded by the two radii v and $v + dv$, i.e., by integrating over a shell of radius v and thickness dv. The appropriate volume element for the integration is simply the volume of this spherical shell, which is $4\pi v^2dv$. We substitute this for the volume element $dv_xdv_ydv_z$ in the above expression to give the final form for the Maxwell–Boltzmann distribution of molecular speeds.

$$P(v)dv = 4\pi \left(\frac{m}{2\pi k_BT}\right)^{3/2} v^2 \exp\left(-\frac{mv^2}{2k_BT}\right)dv \tag{5.25}$$

5.3 Mean speed, most probable speed and root-mean-square speed of the particles in a gas

We can use the Maxwell–Boltzmann distribution to determine the mean speed and the most probable speed of the particles in the gas. Since the probability distribution is normalised, the mean speed is determined from the following integral:

$$\langle v \rangle = \int_0^\infty vP(v)dv \tag{5.26}$$

When we substitute for $P(v)$ using equation (5.25) and carry out the integral, we obtain

$$\langle v \rangle = \left(\frac{8k_BT}{\pi m}\right)^{1/2} \quad \text{Mean speed} \tag{5.27}$$

Table 5.1. The mean speed, most probable speed, root-mean-square speed, and mean relative speed are given for some common gases at 300 K.

Molecule	Mass/g mol^{-1}	$\langle v \rangle$/m s^{-1}	v_{mp}/m s^{-1}	v_{rms}/m s^{-1}	$\langle v_{rel} \rangle$/m s^{-1}
H_2	2	1776	1574	1928	2512
CH_4	16	628	557	682	888
N_2	28	475	421	515	672
Cl_2	71	298	262	324	421

With a little more work, this result may be generalised to give the mean relative speed between two particles of masses m_A and m_B.

$$\langle v_{rel} \rangle = \left(\frac{8k_BT}{\pi\mu} \right)^{1/2} \qquad \text{Mean relative speed} \qquad (5.28)$$

where $\mu = m_A m_B/(m_A + m_B)$ is the reduced mass of the particles.

We can find the most probable speed by maximising the Maxwell–Boltzmann distribution with respect to v (a good exercise for the reader who would like some practice at calculus), giving

$$v_{mp} = \left(\frac{2k_BT}{m} \right)^{1/2} \qquad \text{Most probable speed} \qquad (5.29)$$

Another characteristic speed that is often used is the root-mean-square speed, which we met earlier. By rearranging equation (5.20), we find that this is given by

$$v_{rms} = \left(\frac{3k_BT}{m} \right)^{1/2} \qquad \text{Root-mean-square speed} \qquad (5.30)$$

The mean speed, most probable speed, root-mean-square speed, and mean relative speed are given for some common gases at 300 K in table 5.1.

IOP Concise Physics

An Introduction to the Gas Phase

Claire Vallance

Chapter 6

Collision rates in gases

Collisions are one of the most fundamental processes in chemistry, and provide the mechanism by which both chemical reactions and energy transfer occur within a gas. The rate at which collisions occur determines the timescale of these events, and is therefore an important property for us to be able to calculate. The rate of collisions is usually expressed as a collision frequency, defined as the number of collisions occurring within a given volume per unit time. We will use kinetic theory to calculate collision frequencies for two cases: collisions of gas particles with the container walls; and collisions between the gas particles themselves.

6.1 Collisions with the container walls

We have already completed much of the work required to calculate the frequency of collisions with the container walls. In section 5.1, we showed that for a wall of area A, all molecules in a volume $A v_x \Delta t$ with positive velocities will collide with the wall in the time interval Δt. We can use our probability distribution $P(v_x)$ from equation (5.23) to determine the average value $\langle V \rangle$ of this volume.

$$
\begin{aligned}
\langle V \rangle &= A\Delta t \int_0^\infty v_x P(v_x) dv_x \\
&= A\Delta t \left(\frac{m}{2\pi k_B T} \right)^{1/2} \int_0^\infty v_x \exp\left(-\frac{m v_x^2}{2 k_B T} \right) dv_x \\
&= A\Delta t \left(\frac{k_B T}{2\pi m} \right)^{1/2}
\end{aligned}
\tag{6.1}
$$

doi:10.1088/978-1-6817-4692-0ch6

Multiplying the result by the number density of molecules, $\frac{N}{V} = \frac{p}{k_B T}$, yields the number of collisions occurring in the time interval Δt. Per unit time and unit area ($t = 1$ s, $A = 1$ m^2), this yields a collision frequency[1]

$$z_{wall} = \frac{p}{k_B T} \left(\frac{k_B T}{2\pi m} \right)^{1/2}$$

$$= \frac{p}{(2\pi m k_B T)^{1/2}} \tag{6.2}$$

Note that since $\left(\frac{k_B T}{2\pi m} \right)^{1/2}$ is equal to $\langle v \rangle / 4$, where $\langle v \rangle$ is the mean speed defined in equation (5.27), the collision frequency is also sometimes written,

$$z_{wall} = \frac{\langle v \rangle N}{4V} \tag{6.3}$$

To give the reader an idea of a typical collision frequency of gas particles with the container walls, at 1 bar pressure and 298 K, N_2 molecules undergo about 2.9×10^{27} collisions with a 1 m^2 area of wall every second.

6.2 Collisions with other molecules

To determine the number of collisions a molecule undergoes with other molecules per unit time, we need to introduce the concept of the collision cross section, σ. This is defined as the cross sectional area that the centres of two particles must lie within if they are to collide. In the kinetic model, because it is assumed that there are no intermolecular forces, the particles act as 'hard spheres', and a collision only occurs when the centres of two particles are separated by a distance equal to the particle diameter, d. This is shown in figure 6.1. Imagine that we have 'frozen' the motion of all of the particles apart from the darker-coloured particle on the left. We can see that as this particle travels through the gas, it will only collide with other particles

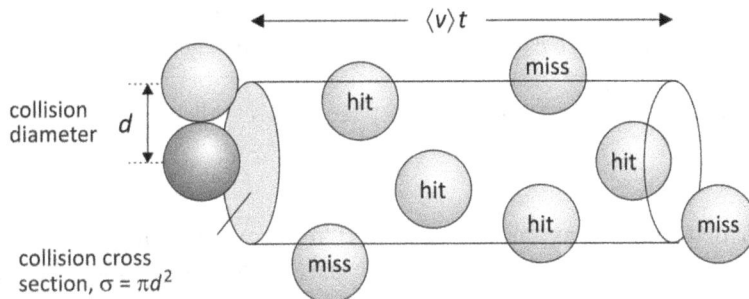

Figure 6.1. A molecule travelling through a gas will collide with all other molecules lying within a cylindrical *collision volume*. The cross-sectional area through this volume is known as the *collision cross-section*.

---

[1] Take care not to confuse the lower-case z used to denote collision frequency with the upper-case Z used to denote the compression ratio.

whose centres lie within the cross-sectional area $\sigma = \pi d^2$. This area defines the collision cross-section.

We can work out the collision frequency by looking at figure 6.1 in a little more detail. Within a time interval Δt, the particle on the left will move a distance $\langle v \rangle \Delta t$ through the gas, represented by the length of the cylinder (as defined previously, $\langle v \rangle$ is the average velocity of the particle). The number of collisions the particle undergoes in the time interval Δt is therefore simply equal to the number of gas particles within the cylindrical *collision volume*. We calculate this number by multiplying the number density of particles in the gas, $N/V = p/(k_B T)$ by the collision volume $\sigma \langle v \rangle \Delta t$ sampled by the particle. We want to know the number of collisions per unit time, so we set $\Delta t = 1$ s. Also, since the other particles are not in fact stationary, we need to replace $\langle v \rangle$, the average speed of one molecule in the gas, by $\langle v_{rel} \rangle$, the mean relative velocity of the gas particles. The collision frequency is therefore:

$$z = \sigma \langle v_{rel} \rangle \frac{N}{V}$$
$$= \sigma \langle v_{rel} \rangle \frac{p}{k_B T} \tag{6.4}$$

For readers familiar with chemical kinetics, note that the process of a single particle colliding with other particles in a gas is a first-order rate process, and that the above expression is in fact simply a first-order rate law of the form $z = $ collision rate $= k[X]$, with k the rate constant and $[X]$ the particle concentration (or in the present case, number density). Comparing this with equation (6.4), we can identify the 'molecular scale' rate constant as the volume of the collision cylinder swept out per unit time, $\sigma \langle v_{rel} \rangle$.

The collision rate z that we have just calculated is the number of collisions experienced by a single molecule per second. Usually, we are not too concerned by the fate of a single molecule within a sea of identical molecules, and a more interesting quantity is the total number of collisions occurring in a given volume per unit time. This quantity is called the *total collision frequency* or *collision density*. Since, on average, all molecules in the gas will undergo the same number of collisions per unit time, we can find the collision density by multiplying the number of collisions undergone by one molecule by the total number of molecules within the volume of interest. For a gas of identical particles, X, this yields

$$z_{XX} = \frac{1}{2} \sigma \langle v \rangle \left(\frac{N}{V} \right)^2 \tag{6.5}$$

The factor of 1/2 in this expression ensures that we avoid double counting of each collision (i.e., the collision of particle X with particle X′ is the same event as that of X′ with X, and should only be counted once, not twice). Substituting for $\langle v_{rel} \rangle$ from equation (5.28), and using the fact that we can relate concentrations and number densities using $[X]N_A = N_X/V$, we can rewrite equation (6.5) as

$$z_{XX} = \sigma \left(\frac{4k_B T}{\pi m} \right)^{1/2} N_A^2 [X]^2 \tag{6.6}$$

Note that in the above we have used the fact that the reduced mass of two identical particles of mass m is $\mu = m^2/(m + m) = m/2$.

As an example, for N_2 gas under standard conditions ($p = 1$ bar, $T = 298$ K), with a collision diameter of 0.28 nm, the collision density is $z_{XX} = 5 \times 10^{34}$ s^{-1} m^{-3}. Equation (6.6) can easily be extended to cover collisions between different types of molecule, yielding

$$z_{XY} = \sigma \left(\frac{8k_B T}{\pi\mu} \right)^{1/2} N_A^2 [X][Y] \tag{6.7}$$

Note that in this case the collision cross section is still $\sigma = \pi d^2$, but the collision diameter d is now given by $d = (d_X + d_Y)/2$, with d_X and d_Y the diameters of the colliding species X and Y.

6.3 Mean free path

The average distance a molecule travels between collisions is called the *mean free path*, usually given the symbol λ. This is a straightforward quantity to calculate now that we know how to determine the collision frequency. To determine the mean free path, we need to know the mean relative speed of the particles, which we can calculate from equation (5.28), and the time between collisions, which is simply the inverse of the collision frequency, z. Since distance = velocity × time, the mean free path is therefore

$$\lambda = \frac{\langle v_{rel} \rangle}{z} \tag{6.8}$$

At standard pressure and temperature, the mean free path is generally of the order of a few tens of nanometres. Since z is proportional to pressure, λ is inversely proportional to pressure; for example, doubling the pressure halves the mean free path.

For the remainder of this chapter, we will consider some of the ways in which the collisional properties of gases can be harnessed to our advantage for a number of applications.

6.4 Effusion and gas leaks

A simple application of some of the concepts we have covered so far in this chapter on collisions is *effusion*, in which a gas at pressure p and temperature T escapes from a higher pressure region into a lower pressure region through a small hole of area a. Effusion occurs when the diameter of the hole is smaller than the mean free path in the gas, so that no collisions occur as the molecules pass through the hole. It is very straightforward to determine the rate of escape of the molecules, dN/dt, since this is simply the rate at which they 'collide with' the hole.

$$\text{effusion rate, } \frac{dN}{dt} = z_{\text{wall}}a$$

$$= \frac{pa}{(2\pi m k_{\text{B}}T)^{1/2}} \tag{6.9}$$

The fact that the rate of effusion is proportional to $1/m^{1/2}$ was originally observed experimentally, and is known as *Graham's law of effusion*.

As the gas leaks out of the container, the pressure decreases, so the rate of effusion is time dependent. The rate of change of pressure with time is

$$\frac{dp}{dt} = \frac{d(N k_{\text{B}}T/V)}{dt}$$

$$= \frac{k_{\text{B}}T}{V}\frac{dN}{dt} \tag{6.10}$$

Substituting for dN/dt from equation (6.9) and rearranging gives

$$\frac{1}{p}dp = -\left(\frac{k_{\text{B}}T}{2\pi m}\right)^{1/2}\frac{a}{V}dt \tag{6.11}$$

which we can integrate to give

$$p = p_0 \exp(-t/\tau) \tag{6.12}$$

with

$$\tau = \left(\frac{2\pi m}{k_{\text{B}}T}\right)^{1/2}\frac{V}{a} \tag{6.13}$$

Equation (6.13) has a number of uses. In log form, the equation becomes $\ln p = \ln p_0 - t/\tau$. Therefore, if we plot $\ln p$ for the gas inside a container against t, we can determine $\ln p_0$ and τ. A measurement of τ provides a simple way of determining the molecular mass, m, as long as the temperature and volume are held constant. Alternatively, if we have a solid sample inside our container, then the measurement of $\ln p_0$ yields the vapour pressure (see section 1.4).

6.5 Molecular beams

State-of-the-art experiments in a number of areas of physics and physical chemistry—for example, high resolution spectroscopy, reaction dynamics, and surface science—employ molecular beams. Using a beam of molecules provides a sample with a well defined velocity distribution, and allows directional properties of chemical or physical processes to be studied. An example is a crossed-molecular-beam experiment, in which two molecular beams are crossed, usually at right angles, as shown in figure 6.2. A chemical reaction occurs in the crossing region, and the speed and angular distribution of one or more of the products is measured. The measured scattering distribution can then be analysed to gain insight into the forces and energetics involved in the transition-state region, providing a direct probe of the

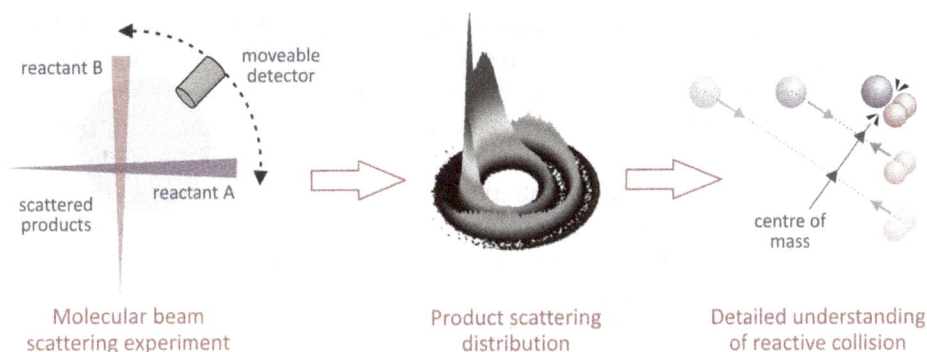

Figure 6.2. A crossed molecular beam experiment allows molecular scattering distributions to be probed in detail.

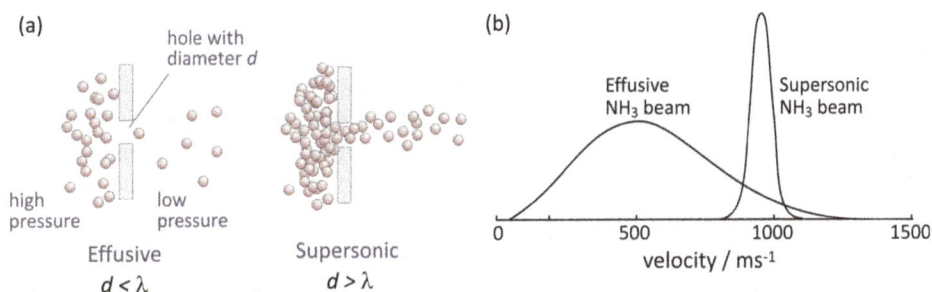

Figure 6.3. (a) Effusive and supersonic molecular beams; (b) The presence or absence of collisions during the expansion is determined by the ratio of orifice diameter d to mean free path inside the source, λ, and has a dramatic effect on the speed distribution within the resulting molecular beam.

fundamental physics underlying chemical reactivity. We can use our knowledge of the behaviour of gases to understand both the formation of a molecular beam and a number of its properties.

There are two types of molecular beam sources, known as *effusive* and *supersonic* sources, respectively. As shown in figure 6.3, both types of source work by allowing gas to escape from a high pressure region through a small orifice into a vacuum. The difference between the two sources is that in an effusive source the diameter of the hole is smaller than the mean free path of molecules in the gas, and in a supersonic source it is larger. The orifice size is generally similar in the two types of source, but the supersonic source operates at a much higher gas pressure, yielding a much shorter mean free path than in the effusive source. The two types of beam have very different properties.

6.5.1 Effusive sources

As the name suggests, effusive sources are based on the phenomenon of effusion described in section 6.4. In an effusive beam, since molecules effectively 'wander' out of the hole whenever they 'collide' with it, the Maxwell–Boltzmann distribution of the molecular speeds in the source is more or less maintained in the molecular beam. In fact, the speed distribution is somewhat skewed towards higher speeds, since faster molecules undergo a greater number of collisions with the walls and are therefore more likely to exit the hole. The velocity components of the molecules within the source are conserved in the molecular beam, with the result that the beam has a broad $\cos^2 \theta$ angular distribution, where θ is the angle between the molecular velocity and the beam axis (i.e., the direction normal to the wall of the chamber containing the hole). Effusive sources generally contain the gas at a low pressure, and are mainly used to produce beams of metal atoms or other species that can only be prepared at low pressure in the gas phase. Usually the source is heated to high temperatures in order to obtain as high a vapour pressure as possible (see section 1.4).

6.5.2 Supersonic sources

In a supersonic source, because the mean free path is much smaller than the diameter of the hole through which the molecules must pass, many collisions occur both as the molecules exit the hole and in the region immediately beyond it. Collisions that impart a velocity component along the beam axis are the most successful at allowing a molecule to escape this region, with the result that the molecules that end up in the beam are those for which the collisions have converted almost all of their random translational energy and internal (rotational and vibrational) energy into directed translational kinetic energy along the beam axis. The beam molecules therefore have almost no internal energy, occupying only very low rotational quantum states, and have a very narrow speed distribution. The angular distribution about the beam axis is also much narrower than for an effusive beam. Since the width of the molecular speed distribution determines the temperature of a gas, by this definition the molecules in a supersonic molecular beam are extremely cold. It is fairly usual to reach temperatures as low as 5 K by this very simple technique of expanding a gas through a small hole. The low temperatures in a molecular beam make them ideal for preparing molecules for spectroscopic studies, since the small number of occupied quantum states often leads to a considerable simplification of the recorded spectrum relative to that of a sample at room temperature.

Chapter 7

Transport properties of gases

As the name suggests, a *transport property* of a substance describes its ability to transport matter, energy, or some other quantity from one location to another. Examples include thermal conductivity (the transport of energy down a temperature gradient), electrical conductivity (transport of charge down a potential gradient), and diffusion (transport of matter down a concentration gradient). We can use the kinetic theory of gases to calculate several transport properties of gases. Firstly, however, we need to introduce the idea of a flux.

7.1 Flux

When dealing with transport properties, we are generally interested in the rate at which matter, energy, charge, or some other property is transported. We usually define this in terms of a flux, which is simply the amount of matter, energy, charge, or other quantity of interest passing through a unit area per unit time. For example, mass flux is measured in units of $kg\ m^{-2}\ s^{-1}$, energy flux is measured in units of $J\ m^{-2}\ s^{-1}$, and so on. As described above, transport of some property occurs in response to a gradient in a related property, and the flux is generally proportional to the gradient. Note that both the flux, J, and the gradient are vector properties. For example, if there is a concentration gradient in some direction z, then there will be a component of mass flux in the same direction.

$$J_z(\text{matter}) \propto \frac{dn}{dz} \qquad (7.1)$$

Here, we are using the shorthand $n = N/V$ to denote the number density[1]. This proportionality between matter flux or *diffusion* and the concentration gradient is

---

[1] Take care not to confuse the number density, n, as defined here, with the number of moles of gas, also denoted by the symbol n.

often referred to as Fick's first law of diffusion. The constant of proportionality is called the diffusion coefficient, and is usually given the symbol D.

$$J_z(\text{matter}) = -D\frac{dn}{dz} \tag{7.2}$$

Note that the negative sign denotes the fact that matter diffuses *down* a concentration gradient from higher to lower concentration. If dn/dz is negative, meaning that the concentration decreases as we move in the positive z direction, then J_z will be positive, i.e., the flow of matter will be in the positive z direction.

Similarly, if there is a temperature gradient dT/dz along z, there will be a component of energy flux along z, which will determine the rate of *thermal diffusion* or *thermal conductivity*.

$$J_z(\text{energy}) = -\kappa\frac{dT}{dz} \tag{7.3}$$

where κ is the *coefficient of thermal conductivity*.

Now that we have defined the various transport phenomena, we will show how the kinetic theory of gases may be used to obtain values for both the diffusion coefficient, D, and the coefficient of thermal conductivity, κ.

7.2 Diffusion

We can use kinetic theory both to explain the molecular origins of Fick's first law of diffusion (that the flux of diffusing molecules is proportional to the concentration gradient) and also to determine a value for the diffusion coefficient, D. To achieve this, we consider the flux of molecules arriving from opposite directions at an imaginary 'window' within a gas, as shown in figure 7.1(a). Within the gas, there is a concentration gradient from right to left, i.e., the concentration decreases from left to right.

Since the motion of the gas molecules is randomised on each collision, the furthest a given molecule is able to travel in a particular direction is on average equal to a distance of one mean free path, λ. This means that to a first approximation we can assume that all of the particles arriving at the imaginary window over a time interval

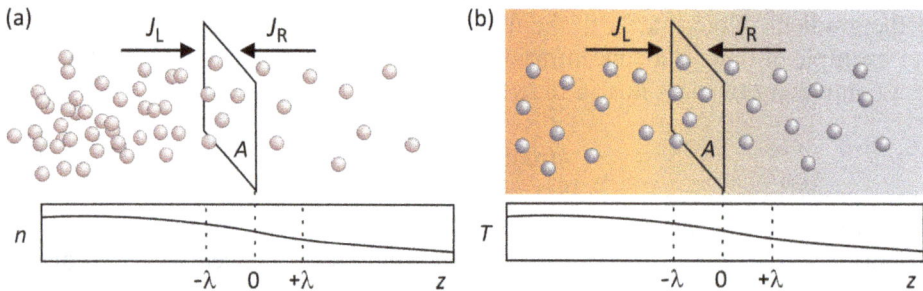

Figure 7.1. Transport properties of gases: (a) diffusion down a concentration gradient; (b) thermal conduction along a temperature gradient. J_L and J_R denote the flux of moleculars and energy, respectively, arriving from the left and right at an imaginary 'window' within the gas.

Δt have arrived there from a distance λ to the left or right, and the number densities of particles arriving from the left and right will therefore reflect the number densities at positions $z = -\lambda$ and $z = +\lambda$, respectively. If we approximate our concentration gradient as being linear between these two points (shown by the two dotted lines on the graph above) with a constant slope equal to the slope at $z = 0$, which we denote $(dn/dz)_0$, then we can use the equation of a straight line to write these two number densities as

$$n(-\lambda) = n(0) - \lambda\left(\frac{dn}{dz}\right)_0$$
$$n(+\lambda) = n(0) + \lambda\left(\frac{dn}{dz}\right)_0 \tag{7.4}$$

From equation (6.3), which gives the number of collisions within a unit area per unit time, the fluxes from the left and right, J_L and J_R, are

$$J_L = \frac{\langle v \rangle n(-\lambda)}{4} = \frac{\langle v \rangle}{4}\left[n(0) - \lambda\left(\frac{dn}{dz}\right)_0\right]$$
$$J_R = \frac{\langle v \rangle n(+\lambda)}{4} = \frac{\langle v \rangle}{4}\left[n(0) + \lambda\left(\frac{dn}{dz}\right)_0\right] \tag{7.5}$$

The net flux in the z direction is therefore

$$J_z = J_L - J_R = -\frac{1}{2}\left(\frac{dn}{dz}\right)_0 \lambda \langle v \rangle \tag{7.6}$$

We have shown that the flux is proportional to the concentration gradient, proving Fick's first law. Comparing equation (7.6) with equation (7.2), it would appear that the diffusion coefficient is given by $D = (1/2)\lambda\langle v \rangle$. In actual fact, the approximations we have made in reaching equation (7.6) mean that this is not quite correct. The main reason for this is that within a distance λ from our window, some molecules are lost through collisions, an effect which needs to be corrected for. A more rigorous treatment yields

$$D = \frac{1}{3}\lambda\langle v \rangle \tag{7.7}$$

We can use this result to predict the way in which the rate of diffusion will respond to changes in temperature and pressure. Increasing the temperature will increase $\langle v \rangle$, and therefore increase the diffusion rate, while increasing the pressure will reduce λ, leading to a reduction in the diffusion rate.

7.3 Thermal conductivity

We can derive equation (7.3), and obtain a value for the coefficient of thermal conductivity, κ, using a similar approach to that used above for diffusion. We will again consider the flux of molecules upon an imaginary window from the left and right, but this time we will assume that the gas has a uniform number density (no concentration gradient), but instead has a temperature gradient, with the temperature decreasing from left to right. This is illustrated in figure 7.1(b). We will assume that the average energy of a molecule is $\varepsilon = \alpha k_B T$, where α is the appropriate multiplier given by the equipartition theorem (see appendix 7.4). For example, a monatomic gas has $\alpha = 3/2$ and $\varepsilon = (3/2)k_B T$.

Using similar arguments to those above for diffusion, namely that molecules are on average reaching the window from a distance of one mean free path away, from regions in which their energies are $\varepsilon(-\lambda)$ and $\varepsilon(+\lambda)$, respectively, we obtain for the energy fluxes from left and right

$$J_L = \frac{1}{4}\langle v \rangle n \varepsilon \left(-\lambda\right) = \frac{1}{4}\langle v \rangle n \alpha k_B \left[T - \lambda\left(\frac{dT}{dz}\right)_0\right]$$

$$J_L = \frac{1}{4}\langle v \rangle n \varepsilon \left(+\lambda\right) = \frac{1}{4}\langle v \rangle n \alpha k_B \left[T + \lambda\left(\frac{dT}{dz}\right)_0\right]$$

(7.8)

The net energy flux is therefore

$$J_z = J_L - J_R = -\frac{1}{2}\alpha\lambda\langle v \rangle k_B n \left(\frac{dT}{dz}\right)_0$$

(7.9)

As in the case of diffusion, there is a correction of 2/3 to be made to account for collisional losses, and the actual flux is

$$J_z = -\frac{1}{3}\alpha\lambda\langle v \rangle k_B n \left(\frac{dT}{dz}\right)_0$$

(7.10)

We have shown that the energy flux is proportional to the temperature gradient, and by inspection of equation (7.10), we can readily determine that the coefficient of thermal conductivity is given by

$$\kappa = \frac{1}{3}\alpha\lambda\langle v \rangle k_B n$$

(7.11)

This expression can be simplified slightly if we recognise that, for an ideal gas, the heat capacity at constant volume is given by $C_V = \alpha k_B N_A$. Substituting this into equation (7.11) yields

$$\kappa = \frac{1}{3}\lambda\langle v \rangle C_V [A]$$

(7.12)

where $[A] = n/N_A = N/(N_A V)$ is the molar concentration. Note that because $\lambda \propto 1/p$ and $[A] \propto p$, the thermal conductivity is independent of pressure[2].

---

[2] This is true at all but very low pressures. At extremely low pressures, the mean free path becomes larger than the dimensions of the container, and the container itself starts to influence the distance over which energy may be transported.

7.4 Summary

Our introduction to the gas phase is now complete. The reader has been introduced to gas properties such as temperature, pressure, and volume, and the relationships between these properties have been explained in terms of the behaviour of gases at a molecular level. We have explored a number of models aimed at describing and understanding this behaviour, including the ideal gas model, the kinetic theory of gases, and various corrections to the ideal gas model that account for the non-ideality of real gases. The kinetic theory of gases has been used to investigate the collisional behaviour of gases in some detail, allowing us to determine collision rates and to understand processes such as effusion, supersonic expansion, diffusion, and thermal conductivity.

IOP Concise Physics

An Introduction to the Gas Phase

Claire Vallance

Appendix: the equipartition theorem

The equipartition theorem states that energy is shared equally amongst all energetically accessible degrees of freedom of a system. This is not a particularly surprising result, and can be thought of as another way of saying that a system will generally try to maximise its entropy (i.e., how dispersed or 'spread out' the energy is within the system) by distributing the available energy evenly amongst all accessible modes of motion.

To give a rather contrived example of equipartition of energy, consider a container of gas in which the gas particles are initially stationary. Imagine we now have some mechanism for 'injecting' an amount of energy randomly into the box. The energy will be shared amongst the gas particles, causing them to move about. While you might not realise it, intuitively you know what this motion will look like. For example, you would be very surprised if the particles moved in the way shown in figure A.1(a), and instead you would probably predict something like the random motion shown in figure A.1(b). This is exactly the same result as predicted by the equipartition theorem: the energy is shared out evenly amongst the x, y, and z translational degrees of freedom.

The equipartition theorem does more than simply predict that the available energy will be shared evenly amongst the accessible modes of motion, and can make quantitative predictions about how much energy will appear in each degree of freedom. Specifically, it states that each quadratic degree of freedom will, on average, possess an energy $\frac{1}{2}k_{B}T$. A 'quadratic degree of freedom' is one for which the energy depends on the square of some property. Some quadratic degrees of freedom include:

1. Translational degrees of freedom have kinetic energy $K_{trans} = \frac{1}{2}mv^2$, with m the mass and v the velocity of the object. The kinetic energy of the object has a quadratic dependence on the object's velocity.

2. Rotational degrees of freedom have kinetic energy $K_{rot} = \frac{1}{2}I\omega^2$, with I the moment of inertia and ω the angular velocity of the object. The kinetic energy has a quadratic dependence on the angular velocity.

doi:10.1088/978-1-6817-4692-0AppA   A-1

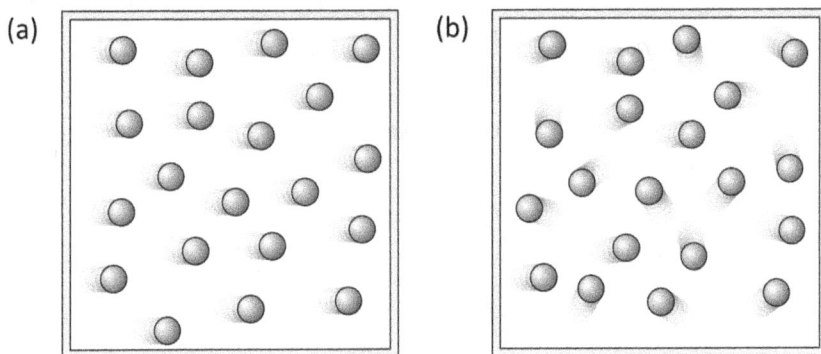

Figure A.1. Motion of particles resulting from an increase in system energy. The energy does not confine itself to one mode of motion, as shown in (a), but is spread out across all available modes of motion, as shown in (b).

3. Vibrational degrees of freedom have kinetic energy $K_{vib} = \frac{1}{2}mv^2$ and potential energy $V_{vib} = \frac{1}{2}kx^2$, with k the force constant of the vibration and x the displacement from the equilibrium atomic positions. The kinetic energy has a quadratic dependence on the velocity of the system, and the potential energy has a quadratic dependence on the displacement from equilibrium.

Note that when considering vibration in a harmonic oscillator potential (V_{vib}, above), we consider both the kinetic energy and the potential energy when counting degrees of freedom. Another important point about vibrations is that vibrational motion in molecules is highly quantised, with large gaps between energy levels relative to the available thermal energy. The classical equipartition theorem breaks down under such circumstances: at room temperature nearly all molecules are present in their ground vibrational state, and higher levels are not thermally accessible. Equipartition contributions from vibrational degrees of freedom usually need only be considered at very high temperatures. Conversely, at room temperature many rotational and translational states are occupied, and these degrees of freedom can be treated classically (i.e., as if their energy levels were not quantised) to a good approximation.

A.1 A simple derivation of the equipartition result for translational motion

To prove that the equipartition principle holds true for translational degrees of freedom, we will use the Maxwell–Boltzmann distribution of molecular speeds (see sections 2.2 and 5.2) to determine the average kinetic energy of a particle in a gas, and will compare this with the predictions of the equipartition principle.

The Maxwell–Boltzmann distribution of molecular speeds in a gas is (see equation (5.25)):

$$P(v) = 4\pi \left(\frac{m}{2\pi k_B T} \right)^{3/2} v^2 \exp\left(-\frac{mv^2}{2k_B T} \right) \tag{A.1}$$

We can average over this distribution to find the average kinetic energy, $\langle K \rangle$ of a particle in the gas.

$$\langle K \rangle = \left\langle \frac{1}{2} mv^2 \right\rangle$$
$$= \int_0^\infty \frac{1}{2} mv^2 P(v) dv \tag{A.2}$$

Substituting for $P(v)$ and taking all constant terms outside the integral gives

$$\langle K \rangle = 2\pi m \left(\frac{m}{2\pi k_B T} \right)^{3/2} \int_0^\infty v^4 \exp\left(-\frac{mv^2}{2k_B T} \right) dv \tag{A.3}$$

We can evaluate the integral by using the general result that

$$\int_0^\infty x^{2s} \exp(-ax^2) = \frac{(2s-1)!!}{2^{s+1} a^s} \left(\frac{\pi}{a} \right)^{1/2} \tag{A.4}$$

where $n!!$ indicates a double factorial, $n(n-2)(n-4)\ldots$ etc. Identifying $x = v$ and $a = m/(2k_B T)$ yields

$$\int_0^\infty v^4 \exp\left(-\frac{mv^2}{2k_B T} \right) dv = \frac{3!!}{8(m/2k_B T)^2} \left(\frac{2\pi k_B T}{m} \right)^{1/2}$$
$$= \frac{3}{2} \left(\frac{k_B T}{m} \right)^2 \left(\frac{2\pi k_B T}{m} \right)^{1/2} \tag{A.5}$$

Substituting this back into our expression for $\langle K \rangle$ gives

$$\langle K \rangle = 2\pi m \left(\frac{m}{2\pi k_B T} \right)^{3/2} \frac{3}{2} \left(\frac{k_B T}{m} \right)^2 \left(\frac{2\pi k_B T}{m} \right)^{1/2}$$
$$= \frac{3}{2} k_B T \tag{A.6}$$

The average translational kinetic energy of a particle in a gas is therefore $\frac{3}{2} k_B T$, or $\frac{1}{2} k_B T$ per translational degree of freedom, in agreement with the equipartition principle.

A.2 A more general derivation of the equipartition theorem

A general derivation of the equipartition theorem requires recourse to statistical mechanics. The following treatment attempts to provide enough explanation to enable readers who are not familiar with the principles of statistical mechanics to understand the derivation.

The partition function in statistical mechanics tells us the number of quantum states of a system that are thermally accessible at a given temperature. It is defined as:

$$q = \sum_i{}' \exp\left(-\frac{E_i}{k_B T}\right) \tag{A.7}$$

where the sum is over the quantum states of the system, and E_i is the energy of quantum state i.

Once the partition function is known, we can calculate many of the macroscopic properties of the system, such as the internal energy, enthalpy, entropy, pressure, and so on, using standard equations from statistical mechanics. Of particular relevance here, to derive the equipartition theorem we will use the partition function to calculate the internal energy U associated with a single degree of freedom of the system. First, however, we need to consider the difference between a quantum and a classical system.

If we intend to treat the particle motions classically, which we should do if we are expecting to derive the classical equipartition theorem, then it does not make much sense to express the partition function as a sum of discrete terms over individual quantum states, as we have above. Classically, the position and momentum of a particle can vary continuously, and the 'energy levels' are therefore also continuous. As a result, the classical partition function takes the form of an integral rather than a sum,

$$q = \int \exp\left(-\frac{E(x_1, x_2 \ldots, p_1, p_2 \ldots)}{k_B T}\right) dx_1 dx_2 \cdots dp_1 dp_2 \cdots \tag{A.8}$$

where the energy is now a function of the particle positions x_i and momenta p_i.

If we assume that we can write the energy as a sum of contributions from each degree of freedom, i.e., $E(x_1, x_2 \ldots, p_1, p_2 \ldots) = E(x_1) + E(x_2) + \cdots + E(p_1) + E(p_2) + \cdots$, then the exponential dependence of the partition function on the energy means that we can separate the integral into the product of integrals over each degree of freedom, i.e.,

$$\exp\left(-\frac{E(x_1, x_2 \ldots, p_1, p_2 \ldots)}{k_B T}\right) = \exp\left(-\frac{E(x_1) + E(x_2) + \cdots + E(p_1) + E(p_2) + \cdots}{k_B T}\right)$$

$$= \exp\left(-\frac{E(x_1)}{k_B T}\right) \exp\left(-\frac{E(x_2)}{k_B T}\right) \cdots \exp\left(=\frac{E(p_1)}{k_B T}\right)$$

$$\exp\left(-\frac{E(p_2)}{k_B T}\right) \cdots$$

and the integral may be written

$$q = \int \exp\left(-\frac{E(x_1)}{k_B T}\right) dx_1 \int \exp\left(-\frac{E(x_2)}{k_B T}\right) dx_2 \cdots$$

$$\times \int \exp\left(-\frac{E(p_1)}{k_B T}\right) dp_1 \int \exp\left(-\frac{E(p_2)}{k_B T}\right) dp_2 \cdots \tag{A.9}$$

$$= q(x_1)q(x_2)\cdots q(p_1)q(p_2)\cdots$$

The consequence of all this algebra is that we have separated the partition function for the system into the product of individual partition functions for each degree of freedom. For a single degree of freedom in which the energy depends quadratically on the coordinate x (i.e., $E(x) = cx^2$ with c a constant), we may write the partition function as

$$q(x) = \int_{-\infty}^{\infty} \exp\left(-\frac{E(x)}{k_B T}\right) dx$$

$$= \int_{-\infty}^{\infty} \exp\left(-\frac{cx^2}{k_B T}\right) dx \tag{A.10}$$

$$= \left(\frac{\pi k_B T}{c}\right)^{1/2}$$

where we have used the standard integral

$$\int_{-\infty}^{\infty} \exp(-ax^2) dx = \left(\frac{\pi}{a}\right)^{1/2} \tag{A.11}$$

From the partition function, we can calculate the internal energy of the system according to the standard result from statistical mechanics:

$$U = k_B T^2 \frac{d \ln q}{dT} \tag{A.12}$$

Substituting the partition function from equation (A.10) into this expression yields the internal energy for a single degree of freedom.

$$U = k_B T^2 \frac{d}{dT} \ln\left(\frac{\pi k_B T}{c}\right)^{1/2}$$

$$= \frac{1}{2} k_B T^2 \frac{d}{dT} \ln\left(\frac{\pi k_B T}{c}\right)$$

$$= \frac{k_B T^2}{2} \frac{c}{\pi k_B T} \frac{\pi k_B}{c} \tag{A.13}$$

$$= \frac{1}{2} k_B T$$

The energy associated with each quadratic degree of freedom is therefore $\frac{1}{2}k_B T$, and we have proved the equipartition theorem.

www.ingramcontent.com/pod-product-compliance
Lightning Source LLC
Chambersburg PA
CBHW082113210326
41599CB00033B/6688